AF429195

After suffering an attack against his exemplary and impeccable professional career, Engineer Israel Laisequilla offers us in this work a detailed and clearly written guide that allows us to delve into the world of industry in the unique manner that only the so-called "most controversial engineer" can achieve.

INSTRUCTIONS

The information presented here takes into account the access and/or availability of current information and technologies. In case you require more information, formats, and/or examples, your inquiry may increase its reliability thanks to the criteria developed with the book. Any additional information, formulas, and/or videos may be requested using your preferred artificial assistant.

the all about Industrial Methods

ENGR'S WORKSHOP

I. LAISEQUILLA

Revision 1

Author's introduction added, July 2024.

Dedicated with much appreciation and respect to all workers in the
manufacturing industry.

the all about Industrial Methods

CONTENT

ACKNOWLEDGEMENTS

This book has been an exciting and challenging adventure, full of ups and downs and moments of creativity and perseverance. But I couldn't have made it this far without the support of some very important people.

First and foremost, I want to thank my family for their love, patience, and constant support at all times. Thank you for believing in me, for inspiring me, and for being my greatest source of motivation and comfort in tough times.

I also want to express my deep appreciation to the editor, who from the beginning showed unwavering faith in my work. Thank you for your ability to understand my vision and bring it to reality, for your professionalism, and dedication to guiding the editing and publishing process of this book.

Last but not least, I want to thank the readers who have supported this book and have been a source of inspiration for me. Thank you for taking the time to read my words, for your feedback and constructive criticism, and for your unwavering support of my work.

This book has been a labor of love, and I hope that, in reading it, you can feel the love and passion I have put into every page. Again, thanks to my family, the editor, and the readers for making this possible. Without you, this book would not exist.

engr's Workshop

INTRODUCTION TO INDUSTRIAL METHODOLOGIES

In today's world, where competition is increasingly intense and globalized, it is crucial for companies to continually improve their production processes and services to stay in the market. Industrial methodologies are a key tool for achieving this continuous improvement.

Industrial methodologies are techniques and tools used to enhance process quality, reduce costs, increase productivity and efficiency, and maximize customer satisfaction. These methodologies are applicable to any industry, from manufacturing to services, and can be employed at any stage of the product or service lifecycle.

An industrial methodology is a structured and systematic set of techniques, tools, and processes applied to enhance the effectiveness and efficiency of a production process or service. These methodologies enable effective resource management, identification of problems and improvement opportunities, and the implementation of solutions for process optimization.

Various industrial methodologies exist, each designed to address specific needs and objectives. Some of the most common ones are described below:

Lean Manufacturing: Focuses on waste elimination and production optimization to improve process efficiency and quality, reducing production time and costs.

Six Sigma: Concentrates on reducing variability and improving quality to achieve stable and predictable processes, eliminating defects and errors in the process.

Kaizen: Based on continuous improvement and waste elimination, aiming to foster a culture of constant improvement by identifying and eliminating sources of waste and improving process quality.

Total Quality Management (TQM): Centers on customer satisfaction and continuous quality improvement, striving for total quality across all areas of the company, from production to customer service.

Business Process Management (BPM): Concentrates on optimizing business processes to identify and eliminate bottlenecks and inefficiencies, improving process efficiency and quality.

Each industrial methodology has specific tools for implementation. For instance, lean manufacturing uses tools like value stream mapping and kanban, while Six Sigma employs tools like root cause analysis and experimental design.

Implementing an industrial methodology is not a simple process; it requires constant commitment and dedication from the entire organization. The company's leadership must clearly define objectives and establish a strategy to achieve them. Employee training in the chosen methodology's tools and techniques is crucial.

Implementing an industrial methodology not only allows for the optimization of production processes but also contributes to the motivation and satisfaction of workers by providing tools and techniques to enhance their performance and contribution to the company. Additionally, the application of these methodologies can contribute to cost reduction and quality improvement, significantly impacting the company's profitability.

Industrial methodologies also enable better risk management in production processes. By identifying and addressing improvement opportunities, the likelihood of errors and process failures is reduced, decreasing the risk of product and service rejections, diminished customer satisfaction, and revenue loss.

Moreover, industrial methodologies can contribute to the environmental and social sustainability of the company. By reducing waste and inefficiency in processes, the consumption of natural and energy resources decreases, positively impacting the environment. Furthermore, improved quality and customer satisfaction can enhance the company's reputation and social responsibility.

In summary, industrial methodologies are a key tool for achieving continuous improvement in a company's production processes and services. These methodologies enable resource optimization, cost reduction, increased productivity and customer satisfaction, risk management, environmental and social sustainability, and improved company profitability.

It is important to note that implementing an industrial methodology is not an isolated process; it requires constant commitment from the entire organization. The company's leadership must clearly define objectives and establish a strategy to achieve them, involving all levels of the organization in the process. Employee training in the tools and techniques of the chosen methodology is essential, as is the continuous measurement and monitoring of results to ensure continuous improvement.

HISTORY OF INDUSTRIAL METHODOLOGIES AND THEIR EVOLUTION

The evolution of industrial methodologies has been a constant throughout history. From the industrial revolution to the present day, companies have sought to improve their processes to make them more efficient and effective. This chapter will describe the evolution of industrial methodologies over time, from early artisanal production systems to modern business management techniques.

Artisanal Production

Before the industrial revolution, most products were crafted in an artisanal manner. Workers were responsible for every stage of the production process, from acquiring raw materials to selling the final product. This form of production was limited in terms of the quantity of products that could be manufactured, and the products varied in quality.

However, artisanal production also had its advantages. Workers had a deep knowledge of the production processes, allowing them to make adjustments and improvements on the fly. Additionally, products were unique and customized, adding value.

The Industrial Revolution and Mass Production

The industrial revolution brought about a radical change in how products were manufactured. Technological advances led to the creation of machines capable of performing repetitive tasks much faster and more efficiently than manual laborers. This marked the beginning of mass production.

Mass production enabled the manufacturing of large quantities of products quickly and efficiently. Furthermore, the products had consistent quality, improving product reliability. However, mass production also had its drawbacks; products became identical, losing their unique and personalized characteristics. Moreover, the production pace was fast, making workers feel like mere cogs in the machine.

Taylorism and Scientific Management

Taylorism is a management methodology developed by Frederick Winslow Taylor in the early 20th century. This methodology was based on the observation and analysis of production processes to enhance efficiency and productivity.

Taylor believed that business management should be a science, and production processes could be studied objectively and analytically. To achieve this, he developed techniques and tools such as time and motion studies, task analysis, process standardization, and worker selection and training.

Taylorism had a significant impact on industry, improving efficiency and productivity in production processes. However, it also faced criticism for turning workers into mere robots, diminishing creativity, and initiative.

Fordism and Assembly Line Production

Fordism is a management methodology developed by Henry Ford in the 1910s. It was based on assembly line production, a system where production tasks were divided into specific and repetitive tasks performed by specialized workers. This approach achieved greater efficiency and reduced production costs.

Fordism had a profound impact on industry, especially in automobile production. Assembly line production allowed the manufacturing of large quantities of vehicles at a lower cost, making automobiles more accessible to the general population.

However, assembly line production had its disadvantages. Workers performed very specific and repetitive tasks, which could be monotonous and boring. Additionally, it required substantial investments in machinery and equipment, making companies highly dependent on market demand.

Total Quality Era and Continuous Improvement

In the 1950s, new business management methodologies emerged, focusing on continuous improvement and total quality. These methodologies were based on the idea that continuous improvement was essential for maintaining competitiveness in the market.

One of the significant methodologies from this period was the Toyota Production System, developed by the Japanese company Toyota. This system emphasized continuous improvement of production processes and waste elimination to enhance efficiency and product quality.

The Toyota Production System had a profound impact on the industry and became a reference for companies worldwide. It also laid the groundwork for the Lean methodology, which focused on waste elimination and continuous improvement.

Total quality was also an important methodology during this time. It emphasized that quality should be a priority in all aspects of the company, from acquiring raw materials to selling the final product. Techniques and tools such as statistical quality control and ISO certification were developed for this purpose.

The Digital Era and Modern Industry

Currently, we are experiencing a new industrial revolution known as Modern Industry. This revolution is based on the digitization and automation of production processes, aiming to improve efficiency and productivity.

Modern Industry relies on technologies such as the Internet of Things, artificial intelligence, and data analysis. These technologies enable connection and communication between different teams and systems, allowing greater coordination and efficiency in production processes.

Moreover, Modern Industry focuses on product customization. Thanks to digitization and automation, it is possible to manufacture customized products efficiently and cost-effectively.

Conclusions

The evolution of industrial methodologies has been a constant throughout history. From artisanal production systems to Modern Industry, companies have sought to improve their processes to become more efficient and effective.

Each methodology described in this chapter has its advantages and disadvantages and has been implemented at different times in history based on the needs of companies and available technological advances.

Currently, Modern Industry is transforming the way products are produced and consumed, with increased efficiency and product customization. However, it also presents new challenges in terms of worker training and adaptation to technological changes.

It is essential to highlight that, beyond specific methodologies, continuous improvement and the pursuit of efficiency in processes have always been fundamental in the industry. This has allowed companies to adapt to market changes and remain competitive over time.

In conclusion, the history of industrial methodologies reflects the evolution of the industry and society as a whole. Each new methodology has been an advancement in terms of efficiency and productivity but has also posed new challenges, requiring adaptation from both companies and workers. The industry continues to advance, and new methodologies and technologies are likely to emerge in the future, but the need for continuous improvement and efficiency will remain a constant in industrial history.

BASIC CONCEPTS OF INDUSTRIAL METHODOLOGIES

Industrial methodologies are a set of techniques and tools used to improve production processes in companies and enhance their efficiency and productivity. These methodologies have been developed over the years to adapt to the needs of each company and sector, and they are based on principles such as continuous improvement, waste elimination, and resource optimization.

This chapter will present the basic concepts of the most common industrial methodologies, such as Lean Manufacturing, Six Sigma, Kaizen, and Total Quality Management (TQM). The fundamental principles of each methodology will be explained, and a comparison will be made to assist managers and professionals in choosing the most suitable methodology for their company.

Lean Manufacturing

Lean Manufacturing is a methodology that focuses on waste elimination and the optimization of production processes. It is based on the concept that any activity that does not add value to the product or service is waste and should be eliminated.

The Lean methodology concentrates on reducing the seven types of wastes: overproduction, waiting time, transportation, unnecessary processes, excessive inventory, unnecessary motion, and defects. To achieve this, tools such as value stream mapping, process standardization, just-in-time (JIT), Kanban system, and continuous improvement are used.

Value stream mapping is a tool that identifies activities that add value and those that do not in the production process. With this information, waste can be eliminated, and the production process can be optimized. Process standardization is used to ensure that all employees follow the same steps in the production process, reducing errors and variability.

Just-in-time (JIT) is a system used to minimize inventory and reduce storage costs. Instead of producing large quantities of products and storing them, only the necessary amount is produced when needed. The Kanban system is used to manage production and inventory, employing a card system to indicate when more units of a product need to be produced.

Continuous improvement is a process involving the constant identification and elimination of waste. This is achieved through the formation of continuous improvement teams, the implementation of a feedback system, and the involvement of all employees in process improvement.

Six Sigma

Six Sigma is a methodology that focuses on reducing variability and eliminating defects in production processes. The goal of Six Sigma is to achieve a process where the number of defects is less than 3.4 per million opportunities.

Six Sigma is based on the DMAIC model (Define, Measure, Analyze, Improve, Control), a systematic approach to process improvement. The first step is defining the problem and establishing project objectives. The next step is measuring process variability and collecting data to identify critical points. A detailed analysis of the data is then conducted to identify the root causes of problems. With this information, solutions can be designed to improve the process. Finally, controls are established to ensure that the process maintains the desired quality level.

In Six Sigma, tools such as the Ishikawa diagram, FMEA (Failure Mode and Effect Analysis), process capability, and correlation analysis are used to identify and resolve problems in the production process.

Kaizen

Kaizen is a methodology that focuses on continuous improvement and is based on the concept that any process can be improved. Kaizen concentrates on process improvement through small constant changes rather than large changes all at once.

In Kaizen, tools such as process flow analysis, standardized work, improvement teams, and waste elimination are used to enhance processes. Process flow analysis identifies activities that do not add value and eliminates them. Standardized work ensures that all employees follow the same steps in the production process.

Improvement teams are groups of employees that regularly meet to identify and solve problems in production processes. Waste elimination is a constant practice in Kaizen that focuses on removing any activity that does not add value.

Total Quality Management (TQM)

Total Quality Management (TQM) is a methodology that focuses on quality in all aspects of the company, not just in production. TQM is based on the concept that quality is the responsibility of all employees, not just production workers.

In TQM, tools such as quality planning, quality control, continuous improvement, and customer satisfaction are used to improve quality throughout the company. Quality planning focuses on establishing quality objectives and developing a plan to achieve them. Quality control is used to ensure that products or services meet established quality standards.

Continuous improvement focuses on identifying and eliminating waste in all aspects of the company. Customer satisfaction focuses on ensuring that products or services meet customer expectations.

Comparison of Methodologies

Each of the mentioned industrial methodologies has its strengths and weaknesses. Lean Manufacturing focuses on waste elimination and process optimization but may not be suitable for companies producing highly customized products. Six Sigma focuses on reducing variability and eliminating defects but may be expensive to implement.

Kaizen focuses on continuous improvement through small constant changes, making it suitable for companies seeking constant improvement rather than a one-time change. TQM focuses on quality in all aspects of the company, making it suitable for companies looking to improve quality in all aspects of their operation.

In terms of implementation, Lean Manufacturing and Kaizen are relatively easy to implement as they focus on small changes and constant improvements in the production process. Six Sigma and TQM, on the other hand, require a greater investment in time and resources for implementation.

Regarding results, Lean Manufacturing and Kaizen tend to produce quick and tangible results in terms of waste reduction and efficiency improvement. Six Sigma and TQM may take more time to produce tangible results but can lead to significant improvements in product or service quality.

It is important to note that none of these methodologies is a one-size-fits-all solution for all companies and situations. Each company should analyze its specific needs and choose the methodology that best suits those needs.

Conclusions

In summary, industrial methodologies are systematic and structured approaches used to improve efficiency and quality in production. There are various industrial methodologies, including Lean Manufacturing, Six Sigma, Kaizen, and TQM, each with its strengths and weaknesses.

The goal of Lean Manufacturing is to eliminate waste in the production process and optimize processes. Six Sigma focuses on reducing variability and eliminating defects in the production process. Kaizen focuses on continuous improvement through small constant changes, while TQM focuses on quality in all aspects of the company.

Each company should analyze its specific needs and choose the methodology that best suits those needs. It is also important to consider that implementing these methodologies requires a significant investment in time and resources, and results may take time to materialize. However, once implemented, these methodologies can lead to significant improvements in production efficiency and quality.

IMPORTANCE OF INDUSTRIAL METHODOLOGIES IN THE CONTINUOUS IMPROVEMENT OF PROCESSES

The deafening roar of machines filled the production floor as workers moved with skill, performing their daily tasks in a well-coordinated dance. However, despite the constant hum of industrial activity, something did not seem to be working quite right.

Such was the case for ABC Company, an electronics manufacturing factory that had been experiencing issues with its production line. Delivery delays and inconsistent product quality had led the company to lose customers and suffer a decline in revenue. In search of a solution, ABC's management team decided to implement industrial methodologies into their production process.

Continuous process improvement through industrial methodologies has been a valuable tool for companies worldwide. These methodologies enable companies to optimize their production processes and improve efficiency in their supply chain, resulting in increased customer satisfaction and higher profits. In this chapter, we will explore the importance of industrial methodologies in continuous process improvement and how they can be successfully implemented in a company.

What are industrial methodologies?

Industrial methodologies are a set of techniques and tools used to improve production processes within a company. These methodologies are designed to help companies optimize their operations and improve the quality of their products, leading to higher customer satisfaction and increased profits.

There are several popular industrial methodologies, each with its strengths and weaknesses. Some common industrial methodologies include:

Lean Manufacturing: Focuses on eliminating any activity that does not add value to the production process, allowing companies to reduce costs and improve product quality.

Six Sigma: Concentrates on reducing variability in production processes, helping companies enhance product quality and reduce costs associated with defects.

Theory of Constraints: Focuses on identifying bottlenecks in the production process and implementing solutions to eliminate them, allowing companies to improve efficiency in their supply chain.

Total Productive Maintenance: Concentrates on improving the maintenance of production equipment, enabling companies to reduce downtime and enhance efficiency in their production process.

5S: Focuses on organizing and cleaning the workspace, helping companies improve workplace safety and efficiency in the production process.

The choice of the appropriate industrial methodology depends on the specific goals of the company and the areas it wants to improve in its production process.

The importance of industrial methodologies in continuous process improvement

Industrial methodologies are essential for continuous process improvement in a company. By implementing these methodologies, companies can improve the quality of their products, reduce costs, and enhance efficiency in their supply chain. This, in turn, can lead to higher customer satisfaction and increased profits.

Additionally, industrial methodologies also help companies identify and resolve issues in their production processes more quickly and efficiently. This is particularly important in an increasingly competitive business environment where speed and efficiency are key to success.

Another benefit of industrial methodologies is that they promote collaboration and teamwork among company employees. By working together to identify and solve problems in the production process, employees can develop a deeper understanding of processes and improve their ability to work together effectively.

How to implement industrial methodologies in a company

Implementing industrial methodologies in a company can be a complex process, but there are some general guidelines that companies can follow to ensure the success of their implementation:

Identify Objectives: Clearly identify the goals the company aims to achieve by implementing an industrial methodology. Is it looking to reduce costs, improve product quality, or enhance efficiency in the supply chain? Once the objectives are identified, it becomes easier to select the appropriate industrial methodology to achieve them.

Training: Ensure that company employees receive proper training on the industrial methodology being implemented. This allows them to understand the methodology, its benefits, and how they can contribute to its success.

Communication: Clearly communicate the implementation of the industrial methodology to all employees. This helps them understand why the methodology is being implemented, what is expected of them, and how they can contribute to its success.

Problem Identification and Resolution: Identify problems in the production process and resolve them before implementing the industrial methodology. This ensures that the implementation of the methodology is more effective.

Monitoring and Evaluation: Monitor and evaluate the implementation of the industrial methodology to ensure that desired objectives are being achieved. This allows the company to make adjustments if necessary and ensure that the methodology is being implemented effectively.

Conclusion

Industrial methodologies are essential for continuous process improvement in a company. By implementing these methodologies, companies can improve product quality, reduce costs, and enhance efficiency in their supply chain. Moreover, industrial methodologies help companies identify and resolve issues in their production processes more quickly and efficiently. By following some general guidelines, companies can successfully implement industrial methodologies and improve their production process effectively.

TYPES OF INDUSTRIAL METHODOLOGIES AND THEIR APPLICATIONS

The industry has evolved significantly in recent years, giving rise to a variety of methodologies aimed at improving the efficiency and quality of industrial processes. In this chapter, we will explore some of the most widely used methodologies in the industry, along with their applications and benefits.

Industrial Methodologies:

Lean Manufacturing:

Lean Manufacturing is a methodology that focuses on reducing production times and eliminating waste in the process. It is based on eliminating any activity that does not add value to the final product or service. Some of the tools used in this methodology include Kanban, Just-In-Time (JIT) production, and continuous improvement.

One of the most well-known applications of Lean Manufacturing is in automobile manufacturing. Toyota was one of the first companies to adopt this methodology, enabling them to improve the quality and efficiency of their production. Today, many companies use this methodology to optimize their processes and improve profitability.

Six Sigma:

Six Sigma is another methodology that seeks to improve the quality of industrial processes. It focuses on eliminating defects in processes and reducing variability. The goal of Six Sigma is to achieve nearly perfect quality in production.

Six Sigma uses a statistical methodology to measure quality and reduce defects in processes. It is divided into five phases: Define, Measure, Analyze, Improve, and Control (DMAIC). Some tools used in Six Sigma include control charts, Pareto analysis, and linear regression.

Six Sigma has been successfully applied in the food industry, medical device manufacturing, and electronic component production. Companies like General Electric and Motorola have implemented this methodology in their processes, improving product quality and reducing production costs.

Total Productive Maintenance (TPM):

Total Productive Maintenance focuses on improving the efficiency of industrial equipment. It is based on preventing equipment failures and continuously improving them to reduce downtime and enhance productivity.

TPM consists of eight pillars: Focused improvement, Autonomous maintenance, Planned maintenance, Training and education, Quality maintenance, Safety maintenance, Administrative maintenance, and Continuous improvement. Each of these pillars focuses on a specific area of equipment maintenance.

TPM has been successfully used in manufacturing, food production, and the chemical industry. Companies like Coca-Cola and Toyota have implemented TPM in their processes, improving equipment efficiency and reducing maintenance costs.

Quick Response Manufacturing (QRM):

Quick Response Manufacturing focuses on reducing production response times and eliminating wait times in the process. It is based on eliminating production bottlenecks and reducing tool change times. Tools used in QRM include value stream analysis and capacity management.

QRM has been successfully used in manufacturing, food production, and the healthcare industry. Companies like Harley-Davidson and L'Oréal have implemented QRM in their processes, reducing product delivery times and improving customer satisfaction.

Design for Six Sigma (DFSS):

Design for Six Sigma focuses on improving the quality of product design by identifying and eliminating design defects. Tools used in DFSS include risk analysis and customer voice evaluation.

DFSS has been successfully used in the automotive industry, medical device production, and the electronics industry. Companies like Ford and Philips have implemented DFSS in their design processes, improving product quality and reducing production costs.

Theory of Constraints (TOC):

The Theory of Constraints focuses on identifying and eliminating production bottlenecks. It involves identifying the most critical constraint in the process and eliminating impediments that limit production. Tools used in TOC include critical chain analysis and inventory management.

TOC has been successfully used in manufacturing, food production, and the healthcare industry. Companies like Procter & Gamble and Eli Lilly have implemented TOC in their processes, identifying production bottlenecks and improving process efficiency.

Conclusion:

The implementation of these methodologies in the industry can lead to a significant improvement in process efficiency and quality, resulting in increased profitability and customer satisfaction. Each methodology focuses on a specific area of improvement, but they all share the common goal of pursuing excellence in production.

It is important to note that there is no perfect methodology, and each company must evaluate which is most suitable for its processes and objectives. Implementing these methodologies requires a change in corporate culture, investment in training, and active participation from all employees.

Ultimately, the application of these methodologies in the industry can be the key to improving process efficiency and quality, translating into greater profitability and customer satisfaction. It is crucial for companies to be willing to invest in the implementation of these methodologies and in the training of their personnel to ensure long-term success.

PROCESSES AND PROCEDURES OF INDUSTRIAL METHODOLOGIES

In the industrial world, efficiency and productivity are fundamental to achieving success. To achieve this, various methodologies have been developed to improve the processes and procedures of companies, with the aim of optimizing the use of resources and reducing costs. In this chapter, we will explore the main industrial methodologies and how they can be applied in different business contexts.

Optimizing processes and procedures can be a complex challenge for companies. In a constantly changing environment, it is important to have tools and strategies that allow quick adaptation to new market demands and maintain a competitive position. Below, we will look at the most popular methodologies and how they are applied in the industry.

Lean Manufacturing Methodology

The Lean Manufacturing methodology, also known as lean production, is a technique that focuses on eliminating anything that does not add value to the production process. This technique originated in Japan in the 1950s, thanks to Toyota and its production system. The main goal of this methodology is to reduce costs, increase efficiency, and improve product quality.

The Lean Manufacturing methodology is based on five fundamental principles:

Identify value: The first step is to determine the value offered to the customer and how this value can be improved.

Map the value stream: Once the value is identified, the production process should be analyzed to identify activities that add value and those that do not.

Create a continuous flow: The goal is to eliminate activities that do not add value and create a continuous workflow.

Establish pull production: Instead of producing based on demand, production is based on actual consumption, avoiding inventory accumulation.

Strive for perfection: The ultimate goal is the continuous improvement of the process by eliminating activities that do not add value and improving those that do.

The application of the Lean Manufacturing methodology can be beneficial for companies, as it allows them to reduce costs, improve efficiency, and increase product quality. Additionally, this methodology can be applied in any type of company, regardless of its size or sector.

Six Sigma Methodology

The Six Sigma methodology is a technique that aims to improve the quality of production processes and reduce errors or defects in products. This methodology is based on a systematic and rigorous approach to identify and correct problems in the production process.

The main goal of the Six Sigma methodology is to reduce variation in the production process, thereby achieving a more consistent and higher-quality final product. To achieve this, two quality levels are established:

Sigma 6: The goal is to reduce defects by 99.99966%, translating to only 3.4 defects per million products.

Sigma 3: The goal is to reduce defects by 93.32%, translating to 66.800 defects per million products.

The Six Sigma methodology is based on five phases:

Define: In this phase, objectives and project scopes are established, customers are identified, and process requirements are set.

Measure: In this phase, the variation of the production process is measured, and quality indicators are established.

Analyze: In this phase, collected data is analyzed, and causes of problems or defects are identified.

Improve: In this phase, solutions are implemented to improve the production process, and tests are conducted to assess their effectiveness.

Control: In this phase, measures are established to maintain implemented changes, and the production process is monitored to ensure it stays within established quality standards.

The Six Sigma methodology can be beneficial for companies as it allows them to reduce errors and defects in products, improve quality, and reduce costs. Additionally, it focuses on the customer and satisfying their needs, which can enhance the company's reputation and increase customer loyalty.

5S Methodology

The 5S methodology is a technique that focuses on the organization and cleanliness of workspaces. This technique was developed in Japan by Toyota as part of its production system.

The 5S methodology is based on five principles:

Sort: Identify and separate necessary elements from unnecessary ones in the workspace.

Set in order: Once necessary elements are identified, organize them so they are easy to find and use.

Shine: Maintain a clean and organized workspace to avoid the accumulation of waste and dirt.

Standardize: Establish procedures and standards to consistently maintain a clean and organized workspace.

Sustain: Consistently maintain a clean and organized workspace by implementing procedures and standards.

The 5S methodology can be beneficial for companies as it improves the organization and cleanliness of the workspace, potentially increasing efficiency and productivity. Additionally, it can enhance workplace safety by reducing the risks of accidents and injuries.

Conclusion

In conclusion, industrial methodologies are valuable tools for improving the processes and procedures of companies. The Lean Manufacturing methodology focuses on eliminating activities that do not add value and creating a continuous workflow, reducing costs, and improving efficiency. The Six Sigma methodology focuses on improving quality and reducing errors and defects in products, increasing customer satisfaction, and enhancing the company's reputation. The 5S methodology focuses on organizing and cleaning the workspace, potentially increasing efficiency and workplace safety. Each of these methodologies can be applied in different business contexts, regardless of the size or sector of the company. It's important to remember that the implementation of these methodologies requires commitment and dedication from both the company and its employees, but the benefits can be significant in terms of efficiency, quality, and profitability.

It is recommended that companies assess their specific needs and objectives before choosing an industrial methodology to implement. Additionally, providing training and support to employees is crucial for successful implementation.

Lastly, it is important to remember that industrial methodologies are not magical solutions to solve all of a company's problems. These tools are useful but must be implemented strategically and adapted to the specific needs and objectives of the company. The key to success lies in constant commitment and dedication to continuous improvement.

In summary, industrial methodologies are a set of techniques and tools used to enhance the processes and procedures of businesses. The Lean Manufacturing methodology concentrates on eliminating non-value-adding activities and establishing a continuous workflow. The Six Sigma methodology centers on enhancing quality, reducing errors and defects in products, and aiming for customer satisfaction. The 5S methodology emphasizes organizing and cleaning the workspace, potentially improving efficiency and workplace safety. Each of these methodologies is applicable across various business domains and can contribute to enhancing efficiency, quality, and profitability. The successful implementation of these methodologies requires commitment and ongoing dedication from both the company and its employees, tailored to the specific needs and goals of the enterprise.

TOOLS AND TECHNIQUES OF INDUSTRIAL METHODOLOGIES

Industrial methodologies are tools and techniques that companies can use to improve the quality, efficiency, and profitability of their production processes. These methodologies are based on the identification and elimination of waste, problems, and errors in production processes.

Industrial methodologies consist of a set of tools and techniques that companies can use to enhance their production processes and increase profitability. Some of the most common industrial methodologies include Lean Manufacturing, Six Sigma, Total Quality Management, Just-in-Time (JIT), Poka-yoke, and Kaizen. Each methodology focuses on specific aspects of production, but they all share the common goal of improving the quality, efficiency, and profitability of processes.

Lean Manufacturing

Lean Manufacturing is a methodology that focuses on eliminating waste in production. Waste includes anything that does not add value to the product or the production process. Waste can include wait times, unnecessary movement, overproduction, defects, excess inventory, among others.

The Lean Manufacturing process involves identifying and eliminating waste in production. Workers use tools and techniques such as Value Stream Mapping and Process Flow Analysis to identify waste in production and design more efficient processes.

Lean Manufacturing is especially useful for companies that produce large quantities of similar products. By eliminating waste in production, companies can reduce costs and improve process efficiency.

Six Sigma

Six Sigma is a methodology used to improve the quality of products and production processes. It is based on the idea that errors and problems in production processes can be identified and eliminated through data analysis and the implementation of solutions.

The Six Sigma process involves the definition, measurement, analysis, improvement, and control (DMAIC) of production processes. Workers use tools and techniques such as Root Cause Analysis and Design of Experiments to identify problems in production and design solutions to improve the quality and efficiency of processes.

Six Sigma is especially useful for companies that produce high-quality products requiring a precise and well-defined production process. By improving product and production process quality, companies can increase customer satisfaction and enhance their market reputation.

Total Quality Management (TQM)

Total Quality Management (TQM) is a methodology used to improve the quality of products and production processes by implementing a quality focus throughout the entire company. It is based on the idea that quality is the responsibility of all workers and departments in the company.

The TQM process involves identifying and eliminating problems and waste in all company processes. Workers use tools and techniques such as Pareto Analysis and Ishikawa Diagram to identify problems and design solutions to improve quality and process efficiency.

TQM is especially useful for companies aiming to improve the quality of their products and production processes across all departments and processes. By implementing a quality focus throughout the company, companies can improve customer satisfaction and the company's reputation.

Just-in-Time (JIT)

Just-in-Time (JIT) is a methodology used to improve the efficiency of production processes by delivering materials, parts, and components just when they are needed for production. It is based on the idea that excess inventory and overproduction are wastes that can be eliminated through just-in-time delivery.

The JIT process involves identifying and eliminating waste in production through just-in-time delivery of materials, parts, and components. Workers use tools and techniques such as Kanban and Continuous Flow to design more efficient processes and reduce wait times and production costs.

JIT is especially useful for companies producing products with variable demand. By reducing excess inventory and overproduction, companies can lower costs and improve production process efficiency.

Poka-yoke

Poka-yoke is a methodology used to prevent errors and problems in production by implementing error prevention devices and systems. It is based on the idea that errors in production are inevitable but can be prevented through the implementation of error prevention devices and systems.

The Poka-yoke process involves identifying errors and problems in production and designing and implementing error prevention devices and systems. Workers use tools and techniques such as Poka-yoke Design to design more efficient processes and reduce errors and problems in production.

Poka-yoke is especially useful for companies producing complex and high-quality products requiring a precise and well-defined production process. By preventing errors and problems in production, companies can improve product quality and reduce production costs.

Kaizen

Kaizen is a methodology used to achieve continuous improvement in production processes through the implementation of small improvements in processes. It is based on the idea that continuous improvement is a constant process, and small improvements in production processes can add up to significant improvements over time.

The Kaizen process involves identifying and eliminating waste and problems in production through the implementation of small improvements in processes. Workers use tools and techniques such as Kaizen Blitz and Value Analysis to design more efficient processes and reduce waste and problems in production.

Kaizen is especially useful for companies seeking continuous improvement in their production processes over time. By implementing small improvements in production processes, companies can achieve significant improvements in the quality, efficiency, and profitability of their production processes.

Conclusion

In summary, industrial methodologies are tools and techniques that companies can use to improve the quality, efficiency, and profitability of their production processes. The implementation of these methodologies can help companies reduce production costs, improve product quality, and increase customer satisfaction.

The six methodologies described in this chapter are just some of the many tools and techniques available for improving production processes in companies. Each company should assess its own needs and goals and select the tools and techniques that best suit its requirements.

It is essential to note that the implementation of these methodologies is not a one-time process but should be a continuous improvement process. Companies must be willing to adapt and change their production processes as business needs and goals evolve.

Furthermore, the successful implementation of these methodologies requires commitment from the company's management and collaboration from all employees. All team members must be willing to learn new tools and techniques and work together to implement changes in production processes.

Ultimately, the implementation of these industrial methodologies can help companies improve the quality of their products, reduce production costs, and enhance the efficiency of their production processes. This can lead to increased customer satisfaction, a better company reputation, and long-term business profitability.

CHOOSING THE APPROPRIATE INDUSTRIAL METHODOLOGY FOR EACH PROJECT

Industrial methodology is a set of techniques, tools, and processes used to design, develop, and improve industrial products and processes. The selection of the appropriate methodology is essential for the project's success, as it can impact process efficiency, costs, and the quality of the final product.

In this chapter, five common industrial methodologies will be described: Design for Manufacturability (DFM), Concurrent Engineering, Design for Quality (DFQ), Lean methodology, and Six Sigma methodology. Additionally, examples of how these methodologies can be applied in different industrial projects will be provided.

Design for Manufacturability (DFM)

Design for Manufacturability (DFM) is a methodology that focuses on designing products to be easy and cost-effective to manufacture. This methodology is used to reduce production costs and improve the quality of the final product.

An example of how DFM methodology can be applied is in the production of plastic parts. In this project, DFM methodology can be utilized to design parts that are easy to mold, reducing production time and costs. For instance, the part can be designed in a way that requires a minimal amount of materials and is easy to extract from the mold without causing damage.

Concurrent Engineering

Concurrent Engineering is a methodology used to expedite product development by involving all stakeholders in the design process from the outset. This methodology is employed to enhance process efficiency and reduce development time.

An example of applying Concurrent Engineering methodology is in the design of a new electronic product. In this project, all stakeholders, including engineers, designers, suppliers, and end-users, can be involved in the design process from the beginning. By working together, stakeholders can quickly identify and address issues, reducing development time and improving the quality of the final product.

Design for Quality (DFQ)

Design for Quality (DFQ) is a methodology used to design high-quality products from the outset. This methodology is employed to enhance the quality of the final product and reduce production costs and cycle times.

An example of applying DFQ methodology is in the production of automobiles. In this project, DFQ methodology can be used to design cars that are safe, reliable, and of high quality. Design for Quality can help identify potential issues in the design and prevent quality defects during the production phase. For example, risk analysis tools can be used to identify and mitigate safety risks in the car design.

Lean Methodology

Lean methodology is an approach that focuses on waste elimination and continuous improvement. This methodology is suitable for projects aiming to reduce costs, increase productivity, and improve the quality of the final product.

An example of applying Lean methodology is in food production. In this project, processes that do not add value, such as wait times, unnecessary movements, and material transport, can be identified. Lean techniques like Kaizen (continuous improvement), Just-in-Time (lean production), and Value Stream Mapping can be applied to reduce costs, improve efficiency, and enhance the quality of the final product.

Six Sigma Methodology

Six Sigma methodology is a statistical approach to process improvement used to reduce defects and enhance the quality of the final product. This methodology relies on data collection and analysis to identify and eliminate root causes of issues in the process.

An example of applying Six Sigma methodology is in the production of medical devices. In this project, processes causing defects can be identified, and data can be collected to analyze process performance. Six Sigma methodology utilizes statistical tools to identify root causes of problems and improve the process to reduce defects and enhance the quality of the final product.

Selection of the Appropriate Methodology

Choosing the right methodology depends on the project type and business objectives. Below are some factors to consider when selecting an industrial methodology:

Project Type: The type of project can influence the choice of the appropriate methodology. For example, Lean methodology may be suitable for mass production projects, while Six Sigma methodology may be more suitable for continuous improvement projects.

Business Objectives: Business objectives can influence the choice of the appropriate methodology. If the business objective is to reduce costs, Lean methodology may be more suitable. If the objective is to improve quality, DFQ or Six Sigma methodologies may be more appropriate.

Available Resources: Available resources, including time, budget, and staff expertise, can influence the choice of the appropriate methodology. Some methodologies may require more resources than others, so it is important to consider resource availability before selecting a methodology.

Corporate Culture: Corporate culture can influence the choice of the appropriate methodology. For example, if the corporate culture values continuous improvement, Six Sigma methodology may be more suitable. If the culture values efficiency, Lean methodology may be more appropriate.

Conclusion

In conclusion, selecting the appropriate methodology is crucial for project success. The right industrial methodology can enhance process efficiency, reduce costs, and improve the quality of the final product. When selecting an industrial methodology, it is important to consider project type, business objectives, available resources, and corporate culture. The methodologies described in this chapter, Design for Manufacturability (DFM), Concurrent Engineering, Design for Quality (DFQ), Lean methodology, and Six Sigma, are just a few of the many options available for selecting an appropriate methodology. A careful evaluation of each methodology is essential, considering that each project is unique and may require a combination of different methodologies for optimal results.

Additionally, it is important to note that the selected industrial methodology is not a one-size-fits-all solution. Each project is unique and may require a combination of different methodologies to achieve the best results.

Lastly, the successful implementation of an industrial methodology depends not only on choosing the right methodology but also on proper project planning, execution, and monitoring. Adequate staff training, allocation of appropriate resources, and continuous process evaluation are fundamental to project success.

In summary, selecting the right methodology is essential to improving process efficiency, reducing costs, and enhancing the quality of the final product in industrial projects. The choice of the right methodology depends on project type, business objectives, available resources, and corporate culture. When selecting a methodology, a careful evaluation is necessary, considering that each project is unique and may require a combination of different methodologies for optimal results. Additionally, proper project planning, execution, and monitoring are crucial for the successful implementation of the selected methodology.

IMPLEMENTATION OF INDUSTRIAL METHODOLOGIES IN THE COMPANY

The implementation of industrial methodologies in the company is a key process for improving efficiency and productivity in production. Industrial methodologies are techniques and tools used to enhance processes and reduce costs. Some of the most popular industrial methodologies include Lean Manufacturing, Six Sigma, Theory of Constraints, and Total Productive Maintenance (TPM).

To implement an industrial methodology in the company, several important steps must be followed. First, it is essential to identify the needs and goals of the company. Clear and specific objectives, such as reducing costs, improving product quality, and increasing production efficiency, should be established. Additionally, it is important for all employees to understand the goals of the methodology and how it will be applied in their daily work.

The next step in the implementation of industrial methodologies is to identify and analyze existing business processes. This includes identifying bottlenecks, points of inefficiency, and waste in production. Detailed analyses of each process should be conducted to identify areas for improvement.

Once issues in the processes have been identified, improvements should be designed and implemented. This may involve reorganizing processes, eliminating waste, automating processes, and improving quality. It is important to involve employees in the implementation of improvements to ensure success.

After implementing improvements, it is important to monitor and control processes to ensure that objectives are being achieved. Key performance indicators should be established, and periodic analyses should be conducted to assess the success of improvements. If issues are identified, adjustments and additional improvements should be made.

The implementation of industrial methodologies can be a challenging process, but it can be a powerful tool for improving efficiency and productivity. Careful evaluation of methodologies and selecting the one that best suits the needs and goals of the company is crucial. Employee training and education are key to ensuring the success of the implementation.

Below are some examples of companies that have successfully implemented industrial methodologies:

Toyota: Toyota is one of the most successful companies in implementing Lean Manufacturing. The company has used this methodology to improve production efficiency and reduce costs. Implementing Lean Manufacturing has allowed Toyota to produce more vehicles with fewer resources and improve the quality of its products. The company has also implemented Total Productive Maintenance (TPM) to enhance efficiency and quality in production.

General Electric: General Electric has implemented the Six Sigma methodology throughout its company. This methodology has enabled the company to improve the quality of its products and services, reduce costs, and increase efficiency in production. The implementation of Six Sigma has been key in transforming GE into a more quality- and efficiency-focused company. The company has also implemented the Design for Six Sigma (DFSS) methodology to improve the quality of product and service design.

Nestlé: Nestlé has implemented the Total Productive Maintenance (TPM) methodology in its production operations worldwide. This methodology has allowed the company to reduce costs and improve efficiency in production. Nestlé has also used Lean Manufacturing methodology to eliminate waste in production and improve the quality of its products.

Boeing: Boeing has implemented the Theory of Constraints methodology in its aircraft production. This methodology has enabled the company to identify and eliminate bottlenecks in production, improving efficiency and reducing costs.

Boeing has also used Lean Manufacturing methodology to enhance efficiency in production and reduce lead times.

Amazon: Amazon has implemented Lean Manufacturing methodology in its logistics and storage operations. The company has used this methodology to reduce lead times and improve efficiency in product delivery. Amazon has also used the Six Sigma methodology to improve the quality of its services and reduce errors in product delivery.

In summary, the implementation of industrial methodologies in the company is essential for improving efficiency, reducing costs, and enhancing the quality of products and services. The selection of the appropriate methodology should be based on the needs and goals of the company and should involve all employees to ensure success. Companies that have successfully implemented industrial methodologies have achieved significant improvements in efficiency, quality, and profitability.

EVALUATION AND MONITORING OF THE EFFECTIVENESS OF INDUSTRIAL METHODOLOGIES

The methodology review meetings are an effective tool for monitoring the effectiveness of industrial methodology. During these meetings, Key Performance Indicators (KPIs) are reviewed, and the obtained results are discussed. Additionally, improvement opportunities are identified, and action plans are established to implement changes.

It is important that these meetings are held consistently, involving all team members to ensure that objectives are being met and to foster collaboration in implementing changes.

During these meetings, it is necessary to review and analyze the established KPIs for evaluating the performance of industrial methodology. These KPIs may include aspects such as production time, product quality, machinery performance, and compliance with safety standards. Clear goals should be defined for each KPI.

Moreover, these meetings serve to identify improvement opportunities in processes and the implementation of industrial methodology. Encouraging ideas and suggestions from all team members is essential for identifying these opportunities and establishing action plans to implement changes.

In summary, methodology review meetings are an essential tool for monitoring the effectiveness of industrial methodologies. They should be conducted consistently, reviewing KPIs and identifying improvement opportunities to implement changes and achieve continuous improvement in company processes.

Internal Audits

Internal audits are a valuable tool for assessing compliance with quality standards and identifying improvement opportunities in industrial methodology. These audits are conducted regularly, focusing on evaluating process efficiency, product quality, and compliance with established standards.

Impartiality is crucial in conducting audits, involving all team members to ensure that industrial methodology objectives are met. During audits, all aspects of industrial methodology, from product quality to process efficiency, should be reviewed.

Clear objectives should be set for audits, and action plans should be established to implement changes. Constant monitoring is essential to ensure that changes are implemented and objectives are met.

In summary, internal audits are an essential tool for evaluating compliance with quality standards and identifying improvement opportunities in industrial methodology. They should be conducted regularly, involving all team members, and action plans should be established to implement changes and achieve continuous improvement in company processes.

Customer Satisfaction Surveys

Customer satisfaction surveys are a valuable tool for evaluating the effectiveness of industrial methodology from the customer's perspective. These surveys focus on measuring customer satisfaction with the company's products and services, as well as the efficiency of processes and meeting expectations.

Regularly conducting these surveys and involving all customers is important to ensure that their expectations are met and to identify improvement opportunities in industrial methodology. Clear objectives should be set for surveys, and action plans should be established to implement changes based on the results obtained.

Customer satisfaction surveys should include questions that assess aspects such as product quality, service efficiency, customer service, and meeting expectations. Specific questions should be established for evaluating each of these aspects, with the opportunity for additional comments to obtain detailed and specific information.

In summary, customer satisfaction surveys are an essential tool for evaluating the effectiveness of industrial methodology from the customer's perspective. They should be conducted regularly, involving all customers, and action plans should be established to implement changes based on the results obtained.

Data Analysis

Data analysis is an essential tool for monitoring the effectiveness of industrial methodology. This analysis focuses on collecting and analyzing data related to process performance and product quality to identify patterns and trends.

Consistently collecting data and setting clear objectives for data analysis are important. Involving all team members is crucial to ensure that objectives are met and to foster collaboration in implementing changes.

During data analysis, it is necessary to identify patterns and trends to assess process performance and product quality. Specific objectives should be established for each of these aspects, and changes should be implemented based on the results obtained.

In summary, data analysis is an essential tool for monitoring the effectiveness of industrial methodologies. Consistent data collection, clear objectives, and implementation of changes based on results are crucial.

Implementation of Changes

The implementation of changes is essential for continuous improvement in company processes and ensuring the effectiveness of industrial methodologies. Action plans should be established to implement changes based on the results obtained from industrial methodology review meetings, customer satisfaction surveys, and data analysis.

These action plans must be specific and detailed, including clear objectives, timelines, and responsibilities for each team member involved. Metrics should be established to evaluate the success of implemented changes, and constant monitoring is necessary to ensure that objectives are met.

Collaboration among team members in implementing changes should be encouraged, and opportunities for training and professional development should be provided to ensure that everyone is aligned and committed to the continuous improvement of processes.

In summary, the implementation of changes is essential for continuous improvement in company processes and ensuring the effectiveness of industrial methodologies. Specific and detailed action plans, metrics to evaluate success, and collaboration and professional development of team members are important.

Conclusion

The assessment and monitoring of the effectiveness of industrial methodologies are essential for continuous improvement in company processes, customer satisfaction, and process efficiency. Methodology review meetings, customer satisfaction surveys, data analysis, and the implementation of changes are essential tools for carrying out this assessment and monitoring process.

Clear objectives and specific action plans should be established for each of these tools, and collaboration and professional development of team members should be encouraged to ensure success in implementing changes and continuous improvement of processes.

In conclusion, the assessment and monitoring of the effectiveness of industrial methodologies are continuous processes that require constant commitment and a mindset of continuous improvement from all team members.

COMMUNICATION AND COLLABORATION IN THE IMPLEMENTATION OF INDUSTRIAL METHODOLOGIES

Implementation of industrial methodologies is a process that aims to improve efficiency and productivity in a company. These methodologies involve the application of specific tools and techniques to enhance the company's processes and systems. However, for the implementation of these methodologies to be successful, it is essential to have good communication and collaboration among different departments and teams within the company. This chapter explores the importance of communication and collaboration in the implementation of industrial methodologies.

Communication in the Implementation of Industrial Methodologies:

Communication is a key factor in the implementation of industrial methodologies as it enables different departments and teams within the company to work together to achieve common objectives. Effective communication allows teams to work more efficiently, identify and resolve issues more quickly, and make informed decisions.

In the context of implementing industrial methodologies, communication should be bidirectional. This means that it is not only important for managers and leaders to communicate with employees, but it is also crucial for employees to communicate with managers and leaders. Bidirectional communication allows employees to express their ideas, concerns, and questions, which can help improve the implementation of industrial methodologies.

Additionally, it is important for communication in the implementation of industrial methodologies to be clear and concise. Clear communication helps avoid misunderstandings and reduces the likelihood of errors. It is also important for communication to be in a language that is understandable to all team members, ensuring a common understanding of objectives and processes involved.

Collaboration in the Implementation of Industrial Methodologies:

Collaboration is another key factor in the implementation of industrial methodologies. Collaboration refers to the ability of different departments and teams within the company to work together to achieve common goals. In the context of implementing industrial methodologies, collaboration can help ensure that all team members are aligned with goals and work together to achieve them.

Effective collaboration in the implementation of industrial methodologies involves active participation from all team members. This means that all team members should be aware of their roles and responsibilities and work together to achieve common goals. Additionally, it is important to establish clear and effective communication channels to facilitate collaboration among team members.

Another important aspect of collaboration in the implementation of industrial methodologies is the ability of team members to share information and knowledge. This can help improve the quality of the company's processes and systems and ensure that all team members have a common understanding of the processes and objectives involved.

Benefits of Good Communication and Collaboration in the Implementation of Industrial Methodologies:

Good communication and collaboration in the implementation of industrial methodologies can generate several benefits for the company, including:

Improved Efficiency: When different departments and teams work together and share information, efficiency in the company can be improved. Team members can identify and resolve problems more quickly and efficiently, reducing the time taken to complete a task or project. Additionally, collaboration can help identify areas for improvement in processes, leading to greater efficiency in the future.

Error Reduction: Good communication and collaboration can also help reduce errors. By having a common understanding of processes and objectives, team members can avoid misunderstandings and make informed decisions, leading to a reduction in errors. This can save time and resources in error correction and rework.

Increased Productivity: The implementation of industrial methodologies aims to improve productivity in the company. Good communication and collaboration can help ensure that all team members are working together towards the same goals, thereby enhancing overall productivity. Additionally, when team members work together more efficiently, they can complete more tasks in less time, further improving productivity.

Improved Work Environment: Effective communication and collaboration can enhance the work environment in the company. By working together towards common goals, team members can develop stronger and more collaborative relationships, improving morale and job satisfaction. When employees feel valued and have a positive relationship with their colleagues, they are more likely to be more productive and willing to contribute to the company in the long run.

Quality Improvement: Good communication and collaboration can contribute to improving the quality of the company's processes and systems. By sharing information and knowledge, team members can identify areas for improvement and work together to enhance overall quality. Additionally, when employees are working together more efficiently and effectively, they can ensure that processes and systems are being followed correctly, improving the quality of the company's products and services.

Increased Innovation: Communication and collaboration can foster innovation in the company. By working together and sharing ideas, team members can generate new ideas and solutions to existing problems. Effective collaboration can help identify opportunities for improvement and evolution of existing processes and systems, leading to significant innovations in the company.

Enhanced Customer Satisfaction: Good communication and collaboration can contribute to improving customer satisfaction. By working together to enhance the quality of the company's products and services, team members can ensure that customers receive the best possible products and services. Additionally, when employees are working together more efficiently and effectively, they can ensure that orders are completed on time, further improving customer satisfaction.

Greater Adaptability: Good communication and collaboration can also help companies become more adaptable to changes in the market and industry. By working together and sharing information, team members can identify and respond more quickly to changes in market demand, new trends, and advances in technology. The ability to adapt quickly to these changes can be a competitive advantage for the company.

Improved Leadership: Good communication and collaboration can enhance leadership capabilities in the company. Leaders who encourage collaboration and effective communication can create a positive and collaborative work environment. Additionally, when leaders are involved in communication and collaboration, they can identify opportunities for improvement and lead change in the company.

Time and Resource Savings: Good communication and collaboration can help save time and resources in the company. By working together more efficiently, team members can complete tasks in less time, reducing labor costs. Additionally, by identifying and resolving problems more quickly and effectively, the cost of rework and error correction can be reduced.

Conclusion

In conclusion, communication and collaboration are crucial in the implementation of industrial methodologies. Effective communication allows different departments and teams within the company to work more efficiently, identify and resolve issues more quickly, and make informed decisions. Effective collaboration involves active participation from all team members and the ability to share information and knowledge. Good communication and collaboration can generate a range of benefits for the company, including improved efficiency, error reduction, increased productivity, improved work environment, quality improvement, increased innovation, enhanced customer satisfaction, greater adaptability, improved leadership, and time and resource savings. Therefore, it is important for companies to pay attention to communication and collaboration when implementing industrial methodologies and work to improve these aspects in their organizational culture.

IDENTIFICATION AND RESOLUTION OF PROBLEMS WITH INDUSTRIAL METHODOLOGIES

In industry, problems are an inevitable part of the production process. Often, these issues can impact the quality, efficiency, and profitability of a company. Therefore, it is essential to have an effective method for identifying and resolving problems in an industrial environment.

This chapter discusses industrial methodologies used to identify and solve problems in the industry. These methodologies include Six Sigma, Lean Manufacturing, and Total Quality Management (TQM). Additionally, common stages used in these methodologies will be discussed, along with some tools and techniques used for problem identification and resolution.

Six Sigma

Six Sigma is a methodology used to improve the quality of a company's products and services. It focuses on reducing variation in production processes and enhancing customer satisfaction. Six Sigma employs a data-driven approach to identify and resolve problems.

The Six Sigma methodology is based on a five-stage process known as DMAIC (Define, Measure, Analyze, Improve, and Control). Each stage has specific objectives and uses different tools and techniques to achieve these goals.

The first stage of DMAIC is Define, where the problem is defined, and project objectives are established. Customer and market needs are also identified.

The second stage is Measure, where data on the production process is collected to determine process capability and stability.

The third stage, Analyze, involves analyzing the data collected in the previous stage to identify the root causes of the problem. Statistical tools such as Pareto Analysis and Ishikawa Diagram are used to identify root causes.

The fourth stage, Improve, focuses on implementing solutions to resolve the problem. Tools such as Design of Experiments and Value Engineering are used to enhance the production process.

The final stage, Control, involves implementing control measures to ensure the production process remains stable and capable. Tools such as Control Plan and Control Chart are used to keep the process under control.

Lean Manufacturing

Lean Manufacturing is another methodology used to improve quality and efficiency in production. It concentrates on eliminating waste and enhancing the efficiency of the production process. Lean Manufacturing uses a continuous improvement approach to identify and solve problems.

The Lean Manufacturing methodology is based on a four-stage process known as the Plan-Do-Check-Act (PDCA) Cycle, which includes Planning, Doing, Checking, and Acting. Each stage has specific objectives and employs different tools and techniques to achieve these objectives.

The first stage of the PDCA cycle is Plan, where the problem is defined, and project objectives are established. Performance measures used to gauge project success are identified.

The second stage is Do, where solutions are implemented to solve the problem, using an action-oriented approach.

The third stage, Check, involves collecting data to assess the success of implemented solutions. Tools like Value Stream Mapping and Fishbone Diagram are used to evaluate process performance.

The final stage, Act, involves taking measures to continuously improve the process. Tools like Kaizen and A3 are used to enhance the production process and eliminate waste.

Total Quality Management (TQM)

Total Quality Management is a methodology used to improve quality and efficiency in production. It focuses on enhancing quality in all aspects of the company, including culture, processes, and products. TQM uses a continuous improvement approach to identify and solve problems.

The TQM methodology is based on an eight-stage process known as the Enhanced Deming Cycle (PDCA). The eight stages include Plan, Do, Check, Act, Analyze, Design, Implement, and Maintain. Each stage has specific objectives and uses different tools and techniques to achieve these objectives.

The first stage of the Enhanced Deming Cycle is Plan, where the problem is defined, and project objectives are established. Performance measures used to gauge project success are identified.

The second stage is Do, where solutions are implemented to solve the problem.

The third stage is Check, where data is collected to evaluate the success of implemented solutions.

The fourth stage is Act, where measures are taken to continuously improve the process.

The fifth stage is Analyze, where data collected in the previous stage is analyzed to identify the root causes of the problem.

The sixth stage is Design, where solutions are designed to solve the problem.

The seventh stage is Implement, where the designed solutions are implemented.

The last stage is Maintain, where control measures are implemented to ensure the stability and capability of the production process.

Tools and Techniques for Problem Identification and Resolution

There are several tools and techniques commonly used in industrial methodologies to identify and resolve problems. Some of these tools and techniques include:

Ishikawa Diagram: Also known as a Fishbone Diagram, it is used to identify the root causes of a problem.

Pareto Analysis: Used to identify the most significant problems affecting the quality and efficiency of the production process.

Control Chart: Used to monitor the production process and detect deviations from the process.

Control Plan: Used to establish control measures necessary to ensure the quality and efficiency of the production process.

Value Stream Mapping: Used to analyze the flow of the production process and identify areas of waste.

Kaizen: Refers to continuous improvement and is used to identify opportunities for improvement in the production process.

A3: A problem-solving tool used to document the improvement process and communicate results to stakeholders.

FMEA (Failure Mode and Effect Analysis): Used to identify potential failures in the production process and design control measures to prevent them.

5S: Refers to workplace organization, cleanliness, and standardization to improve the efficiency and safety of the production process.

Kanban: Used to manage inventory and ensure that necessary materials are available when needed.

These are just some of the tools and techniques used in industrial methodologies to identify and resolve problems. Each tool and technique has its own benefits and is used in different situations.

Conclusions

Industrial methodologies are a set of techniques and tools used in the industry to improve the quality and efficiency of the production process. These methodologies are based on the idea of continuous improvement and waste elimination, applied across a wide range of industrial sectors.

The PDCA cycle is one of the most widely used methodologies in the industry, focusing on the continuous improvement of the production process. This methodology is divided into four stages: Plan, Do, Check, and Act. Specific activities are carried out in each stage to identify and solve problems, and control measures are established to ensure the quality of the production process.

Another popular methodology is Total Quality Management, which focuses on continuous improvement in all aspects of the company. This methodology is based on the notion that quality is not only the responsibility of the quality control department but is the responsibility of all employees in the company. To implement Total Quality Management, specific activities such as employee training, waste identification and elimination, and the implementation of control measures to ensure production quality are carried out.

In addition to these methodologies, there are various specific tools and techniques used in the industry to identify and resolve problems. The Ishikawa Diagram is a tool used to identify the root causes of a problem, while Pareto Analysis is used to identify the most important problems that need to be addressed first. The Control Chart is used to monitor the quality of the production process and detect any unwanted variations.

In summary, industrial methodologies are essential to ensure the quality and efficiency of the production process. Specific tools and techniques used in these methodologies, such as the PDCA cycle, Total Quality Management, Ishikawa Diagram, and Pareto Analysis, allow for effective problem identification and resolution, as well as continuous monitoring and improvement of the production process. Managers and engineers should be familiar with these methodologies and tools to apply them effectively and achieve excellence in their production process. The implementation of industrial methodologies can help companies stay competitive and meet the needs and expectations of customers in an increasingly demanding market.

DEVELOPMENT OF ACTION PLANS AND IMPROVEMENT WITH INDUSTRIAL METHODOLOGIES

In any company or industry, the development of action and improvement plans is crucial to achieving sustainable growth and increasing process efficiency. Action and improvement plans consist of a set of strategies and actions implemented to enhance processes, optimize resources, and increase overall productivity.

This chapter will explain how industrial methodologies can be applied to develop effective action and improvement plans. It will also discuss the different stages of the improvement process and how they can be implemented in a company.

Industrial methodologies for developing action and improvement plans

Industrial methodologies comprise a set of techniques and tools used to optimize processes and enhance overall efficiency in a company. These methodologies have been developed over time and successfully applied across various industrial sectors.

Some of the most popular industrial methodologies include Lean Manufacturing, Six Sigma, Total Quality Management (TQM), Kaizen, and the Theory of Constraints (TOC).

Lean Manufacturing

Lean Manufacturing focuses on waste elimination and creating value for the customer. It is based on the principle that all activities not adding value to the customer should be eliminated.

Lean Manufacturing is divided into five stages: value identification, value stream mapping, flow creation, pull production system implementation, and continuous improvement.

The identification of value involves identifying the products or services that are valued by customers and eliminating those that are not. Value stream mapping is the process of visualizing the flow of materials and information through the production process. Creating flow involves eliminating bottlenecks and establishing a continuous production flow. The implementation of a pull production system involves producing products or services only when they are needed. Continuous improvement involves implementing measures to eliminate waste and improve the efficiency of the production process.

Six Sigma

Six Sigma aims to reduce variability in processes and improve product or service quality. It operates on the principle that process variability is the main cause of defects.

Six Sigma is divided into five stages: Define, Measure, Analyze, Improve, and Control (DMAIC). The Define stage involves defining the problem and identifying the project scope. The Measure stage involves collecting data and measuring the process. The Analyze stage involves identifying the root causes of the problem. The Improve stage involves implementing solutions to address the problem. The Control stage involves implementing measures to sustain the improvement.

Total Quality Management (TQM)

TQM focuses on continuous improvement of product or service quality and is based on the principle that quality is everyone's responsibility in the company.

TQM is divided into eight stages: leadership, education and training, employee involvement, customer focus, continuous improvement, measurement and results analysis, supplier management, and teamwork.

The leadership stage involves the commitment of top management to the implementation of continuous improvement. Education and training involve instructing employees in quality improvement techniques. Employee involvement entails the active participation of employees in continuous improvement. Customer focus involves meeting the needs and expectations of the customer. Continuous improvement entails implementing measures to continually enhance the quality of the product or service. Measurement and analysis of results involve monitoring and analyzing outcomes to take improvement actions. Supplier management involves the selection and evaluation of suppliers to ensure the quality of the product or service. Teamwork involves collaboration and cooperation among different departments within the company.

Kaizen

Kaizen concentrates on continuous process improvement and is based on the belief that all employees can contribute to continuous improvement.

Kaizen is divided into three stages: problem identification, problem analysis, and problem solution. Problem identification involves identifying areas that need improvement. Problem analysis involves identifying the root causes of the problem. Problem solution involves implementing measures to address the problem and prevent its recurrence.

Theory of Constraints (TOC)

TOC targets the identification and elimination of limitations in processes, operating on the principle that processes are constrained.

TOC is divided into three stages: identification of the constraint, exploitation of the constraint, and elevation of the constraint. Identifying the constraint involves recognizing the limitation in the process. Exploiting the constraint involves maximizing the use of that limitation. Elevation of the constraint involves eliminating or mitigating the limitation to improve the overall system performance.

Improvement Process Stage

The improvement process is divided into four stages: planning, implementation, monitoring, and evaluation.

Planning

The planning stage involves problem identification, definition of improvement objectives, identification of root causes of the problem, and selection of the improvement methodology.

Problem identification involves identifying areas that need improvement. Definition of improvement objectives involves defining desired outcomes. Identification of root causes of the problem involves identifying underlying causes of the problem. Selection of the improvement methodology involves choosing the most suitable methodology to solve the problem.

Implementation

The implementation stage involves putting into action the solutions identified in the planning stage.

This stage involves taking the necessary actions to improve the process and achieve the defined objectives. It also involves communicating and engaging employees in the implementation of solutions.

Monitoring

The monitoring stage involves tracking and measuring the results of the implemented solutions. This stage includes comparing the results with the objectives defined in the planning stage and identifying potential deviations.

Evaluation

The evaluation stage involves assessing the results and identifying additional improvement opportunities. This stage includes identifying factors that contributed to the success or failure of the implementation and learning lessons.

Continuous Improvement Tools

There are several tools and techniques that can be used for continuous process improvement. Some of the most common tools include:

Flowcharts: a tool used to visualize processes and the sequence of activities.

Pareto Charts: a tool used to identify the most significant problems and primary causes.

Ishikawa or Fishbone Diagrams: a tool used to identify the root causes of a problem.

Value Stream Analysis (VSA): a tool used to identify activities that add value to the process.

Process Maps: a tool used to visualize processes and interactions between different departments.

Cost-Benefit Analysis: a tool used to assess the costs and benefits of proposed solutions.

Conclusions

In conclusion, implementing action and improvement plans using industrial methodologies can help companies continuously improve their processes and products or services. These methodologies focus on problem identification, root cause identification, and solution implementation to improve processes. They also emphasize employee collaboration and commitment to continuous improvement.

It is important for companies to select the appropriate improvement methodology for their needs and commit to implementing the identified solutions. Additionally, it is crucial for companies to monitor and evaluate results to identify additional improvement opportunities and ensure that processes continue to improve continuously. Continuous improvement is an ongoing process that never ends, but it can help companies stay competitive in a changing market and meet the needs and expectations of their customers.

RISK AND OPPORTUNITY ASSESSMENT IN THE IMPLEMENTATION OF INDUSTRIAL METHODOLOGIES

The implementation of industrial methodologies is a common practice in operations management within companies. However, this implementation can involve risks and opportunities that must be assessed to ensure the success of the project. In this chapter, the evaluation of risks and opportunities in the implementation of industrial methodologies will be discussed. It will define what is meant by risk and opportunity, methodologies for risk and opportunity assessment will be explained, and the steps for conducting a proper assessment will be detailed. Additionally, some examples of common risks and opportunities that may arise in the implementation of industrial methodologies will be analyzed.

Definition of Risk and Opportunity:

Before delving into the assessment of risks and opportunities in the implementation of industrial methodologies, it is important to define what is meant by risk and opportunity. Risk is defined as the possibility of an event or situation occurring that negatively affects the project or organization. On the other hand, opportunity is defined as a situation that can be beneficial for the project or organization.

The proper evaluation of risks and opportunities in the implementation of industrial methodologies is crucial to avoid potential problems and maximize the project's benefits. A thorough assessment of risks and opportunities can provide valuable information for decision-making and planning the methodology's implementation.

Methodologies for Risk and Opportunity Assessment:

There are different methodologies for assessing risks and opportunities in the implementation of industrial methodologies. The most common ones are explained below:

SWOT Analysis

The SWOT analysis (Strengths, Weaknesses, Opportunities, Threats) is a strategic assessment tool that allows the identification of internal and external factors that may impact the success of a project. This analysis is divided into four categories:

Strengths: Internal factors that are positive for the project or organization.

Weaknesses: Internal factors that may negatively affect the project or organization.

Opportunities: External factors that can be beneficial for the project or organization.

Threats: External factors that may negatively impact the project or organization.

The SWOT analysis is useful for identifying the strengths and weaknesses of the project and the external opportunities and threats that may arise during the methodology implementation.

Risk and Opportunity Analysis

The risk and opportunity analysis is a methodology that identifies and evaluates potential events that can affect the project. The analysis is divided into two categories: risks and opportunities. Risks are events that can have a negative impact on the project or organization, while opportunities are events that can have a positive impact.

The risk and opportunity analysis is useful for identifying events that can negatively or positively affect the project and taking measures to reduce negative impacts and capitalize on opportunities.

Cost Analysis

Cost analysis is a methodology that evaluates the financial impact of implementing the methodology. The analysis focuses on the direct and indirect costs of the project and the potential gains and savings that can be achieved through methodology implementation.

Cost analysis is useful for assessing the financial viability of the project and determining whether the benefits outweigh the costs.

Steps for Risk and Opportunity Assessment

To carry out a proper assessment of risks and opportunities in the implementation of industrial methodologies, it is necessary to follow some key steps:

Identify potential risks and opportunities: In this stage, all possible risks and opportunities that may arise during the implementation of the methodology should be identified.

Evaluate the likelihood of risks and opportunities: In this stage, assess the likelihood of the identified risks and opportunities occurring. It is essential to classify them based on their probability of occurrence and potential impact on the project.

Identify possible mitigation measures: In this stage, identify possible measures to mitigate risks and capitalize on identified opportunities.

Evaluate the costs and benefits of mitigation measures: In this stage, evaluate the costs and benefits of the identified mitigation measures and determine their financial feasibility.

Implement mitigation measures: In this stage, implement the mitigation measures to reduce the negative impact of identified risks and capitalize on opportunities.

Examples of Risks and Opportunities in the Implementation of Industrial Methodologies

Below are some examples of common risks and opportunities that may arise in the implementation of industrial methodologies:

Risks:

Lack of adequate training: If personnel is not adequately trained in the methodology to be implemented, there may be implementation problems and an increase in project time and costs.

Resistance to change: If employees resist the changes involved in the implementation of the methodology, there may be delays and issues in implementation.

Technical problems: Technical problems may arise during the implementation of the methodology, leading to implementation delays and increased costs.

Opportunities:

Efficiency improvement: The implementation of an industrial methodology can improve the efficiency of organizational processes and reduce costs.

Quality improvement: The implementation of an industrial methodology can enhance the quality of products or services offered by the organization, increasing customer satisfaction and loyalty.

Increased productivity: The implementation of an industrial methodology can increase organizational productivity, allowing for higher production without increased costs.

Conclusions

Risk and opportunity assessment is a key step in the implementation of industrial methodologies. It allows for the identification of potential risks and opportunities, the evaluation of their potential impact, and the determination of necessary mitigation measures. Risk and opportunity assessment also enables the assessment of the financial viability of the project and the determination of whether the benefits outweigh the costs.

It is important to note that the implementation of an industrial methodology can have a significant impact on the organization. Therefore, conducting a proper assessment of risks and opportunities is crucial to ensure project success.

Additionally, it is important for the organization to provide adequate training to personnel and foster a culture of change and continuous improvement to ensure successful implementation of the methodology.

In summary, risk and opportunity assessment is a key step in the implementation of industrial methodologies. It allows for the identification of potential risks and opportunities and the determination of necessary mitigation measures to ensure project success. Proper assessment of risks and opportunities, along with adequate training for personnel, is essential for successful implementation of the methodology.

TRAINING AND EDUCATION IN INDUSTRIAL METHODOLOGIES

Industrial methodologies are techniques and tools used in the industry to improve the efficiency and quality of production processes. These methodologies include Lean Manufacturing, Six Sigma, Just in Time, Theory of Constraints, Total Productive Maintenance, among others.

Training and education in these methodologies are essential for companies to implement them effectively and achieve significant improvements in their production processes. This chapter will discuss the importance of training and education in industrial methodologies, the benefits they offer, and some examples of their application in different industry areas.

Importance of training and education in industrial methodologies

Training and education in industrial methodologies are fundamental to enhance the production processes of companies and increase their competitiveness in the market. The application of these methodologies allows for the optimization of production processes, cost reduction, improvement in product quality, and increased company efficiency.

The implementation of these methodologies requires trained personnel. Training and education in these methodologies enable staff to understand the importance of implementing these techniques and tools and how to apply them effectively in the production process.

Benefits of training and education in industrial methodologies

Training and education in industrial methodologies offer numerous benefits for companies, including the following:

Improved efficiency: The implementation of industrial methodologies allows for the identification of waste areas and process improvements, resulting in increased efficiency in the company.

Cost reduction: Identifying waste areas and improving production processes help reduce production costs.

Enhanced product quality: Implementing industrial methodologies identifies and corrects defect areas in the production process, leading to improved product quality.

Increased productivity: Implementing industrial methodologies optimizes production processes, translating into increased company productivity.

Improved customer service: Implementing industrial methodologies enhances efficiency and quality of customer service, resulting in increased customer satisfaction and loyalty.

Areas of application of industrial methodologies

Industrial methodologies can be applied in various industry areas, including:

Production processes: Industrial methodologies can be applied to improve efficiency, reduce production costs, and enhance product quality.

Maintenance: Industrial methodologies can be applied to machinery and equipment maintenance to reduce downtime and improve efficiency.

Distribution and storage: Industrial methodologies can be applied to improve efficiency in product distribution and storage, reduce delivery times, and enhance inventory management.

Project management: Industrial methodologies can be applied in project management to optimize resources, reduce delivery times, and improve project quality.

Supply chain management: Industrial methodologies can be applied in supply chain management to improve efficiency in supplier management, reduce costs, and enhance product quality.

Examples of application of industrial methodologies

Here are some examples of the application of industrial methodologies in different industry areas:

Lean Manufacturing: Focuses on eliminating waste in the production process. An example of its application would be reducing waiting times between operations in a production process, improving efficiency, and reducing production costs.

Six Sigma: Focuses on identifying and eliminating defects in the production process. An example of its application would be reducing the number of defects in a product, resulting in improved product quality.

Just in Time: Focuses on eliminating inventory and producing based on customer demand. An example of its application would be reducing waiting times in product delivery, improving customer service, and reducing inventory costs.

Theory of Constraints: Focuses on identifying and eliminating bottlenecks in the production process. An example of its application would be identifying a bottleneck in a production line and implementing measures to eliminate it, resulting in improved process efficiency.

Total Productive Maintenance: Focuses on improving machinery and equipment maintenance. An example of its application would be implementing a preventive maintenance program in an industrial plant, resulting in reduced downtime and improved plant efficiency.

Conclusion

Training and education in industrial methodologies are essential for companies to implement these techniques and tools effectively and achieve significant improvements in their production processes. Industrial methodologies offer numerous benefits, including improved efficiency, cost reduction, enhanced product quality, increased productivity, and improved customer service.

Industrial methodologies can be applied in different industry areas, such as production processes, maintenance, distribution and storage, project management, and supply chain management. The application of these methodologies in the industry can contribute significantly to improving the competitiveness of companies in the market.

CHANGE MANAGEMENT IN THE IMPLEMENTATION OF INDUSTRIAL METHODOLOGIES

Change management is a key aspect in the implementation of industrial methodologies, as it involves significant changes in processes, organizational culture, and interpersonal relationships. This chapter will address the main aspects related to change management in the implementation of industrial methodologies, including factors influencing implementation success, strategies for handling resistance to change, roles and responsibilities of change leaders, as well as tools and techniques available to facilitate implementation.

Factors influencing implementation success of industrial methodologies

The implementation of industrial methodologies is a complex process involving multiple factors influencing its success. In general, these factors can be classified into three main categories: organizational culture, change management capability, and planning and execution of implementation.

Organizational culture is one of the most important factors influencing the success of industrial methodology implementation. Organizational culture can be defined as the set of values, beliefs, norms, and behaviors that characterize an organization. A strong and cohesive organizational culture can be a determining factor in the success of industrial methodology implementation. Conversely, a weak or fragmented organizational culture can be a significant obstacle to successful implementation.

Change management capability is another critical factor influencing the success of industrial methodology implementation. Change management refers to the set of processes and activities carried out to plan, implement, and control changes in an organization. Good change management capability is essential to ensure that the implementation of the industrial methodology is carried out effectively and smoothly.

Planning and execution of implementation are also key factors influencing the success of industrial methodology implementation. Careful planning and efficient execution are necessary to ensure that the implementation is carried out effectively and that the desired objectives are achieved.

Strategies for handling resistance to change

Resistance to change is a common phenomenon in the implementation of industrial methodologies. Resistance to change can be a significant barrier to implementation success, as it can impede the change process and lead to employee dissatisfaction. The following are some strategies for handling resistance to change in the implementation of industrial methodologies:

Communicate the benefits of implementation: It is important for employees to understand the benefits of implementing the industrial methodology. Effective communication can help reduce resistance to change and motivate employees to support the implementation.

Involve employees: It is essential to involve employees in the implementation of the industrial methodology. This may include participation in workgroups, identification of problematic areas, and proposing solutions.

Provide training and support: Training and support are essential to help employees adapt to the changes occurring in the implementation of the industrial methodology. Training may include acquiring new skills and tools, while support may include guidance and follow-up to ensure that employees have the assistance they need for successful implementation.

Recognize and reward success: It is important to recognize and reward success in the implementation of the industrial methodology. This may include public recognition, financial incentives, and professional development opportunities. Recognition and rewards can motivate employees to support implementation and overcome resistance to change.

Accept and manage resistance: Finally, it is important to accept that resistance to change is a natural phenomenon in the implementation of industrial methodologies and must be managed effectively. Resistance management may include identifying underlying factors of resistance, taking measures to address these factors, and guiding employees through the change process.

Roles and responsibilities of change leaders

Change leaders play a fundamental role in the implementation of industrial methodologies. Change leaders are individuals who lead and coordinate the change process, ensuring that the implementation of the industrial methodology is carried out effectively and smoothly. The following are some key roles and responsibilities of change leaders in the implementation of industrial methodologies:

Establish a clear vision: Change leaders must establish a clear vision for the implementation of the industrial methodology. This may include defining clear objectives, communicating the vision to all employees, and aligning the vision with organizational values and culture.

Create a sense of urgency: Change leaders must create a sense of urgency around the implementation of the industrial methodology. This may include identifying risks and opportunities associated with non-implementation and communicating these risks and opportunities to all employees.

Develop and maintain a change coalition: Change leaders must develop and maintain a change coalition that supports the implementation of the industrial methodology. This may include identifying key stakeholders in the implementation, creating a project team, and collaborating with other leaders in the organization.

Communicate and educate: Change leaders must communicate and educate employees about the industrial methodology and the need for its implementation. This may include creating communication materials and organizing training and education sessions.

Implement and consolidate change: Finally, change leaders must implement and consolidate the change. This may include identifying obstacles and taking measures to overcome them, monitoring progress, and adapting the implementation as necessary.

Tools and techniques to facilitate the implementation of industrial methodologies

There are several tools and techniques that can be used to facilitate the implementation of industrial methodologies. These tools and techniques can help change leaders and employees plan, execute, and monitor the implementation process. The following are some of the most common tools and techniques used in the implementation of industrial methodologies:

Flowchart: A flowchart is a visual tool used to represent a process in terms of its individual steps and activities. Using a flowchart can help change leaders and employees visualize the implementation process and identify areas for improvement.

SWOT analysis: SWOT analysis is a tool used to assess the strengths, weaknesses, opportunities, and threats of an organization. Using SWOT analysis can help change leaders identify areas that require more attention during the implementation of the industrial methodology.

Gantt charts: Gantt charts are tools used to plan and monitor projects. Using Gantt charts can help change leaders and employees visualize the implementation schedule of the industrial methodology and identify any delays or deviations in the plan.

Responsibility matrix: The responsibility matrix is a tool used to identify the individual responsibilities of team members in a project. Using the responsibility matrix can help change leaders assign specific tasks and responsibilities to team members during the implementation of the industrial methodology.

Surveys and assessments: Surveys and assessments are tools used to collect information and feedback from employees about the implementation of the industrial methodology. Using surveys and assessments can help change leaders evaluate progress and identify any issues or challenges that need to be addressed.

Conclusion

The implementation of industrial methodologies can be a challenging and complex process. However, with proper planning, clear communication, and effective change management, implementation can be successful and can improve the efficiency, quality, and profitability of an organization. Change leaders play a crucial role in the implementation of industrial methodologies and must be willing to take responsibility for leading and coordinating the change process. The use of tools and techniques such as flowcharts, SWOT analysis, Gantt charts, responsibility matrices, surveys, and assessments can help facilitate the implementation of the industrial methodology and identify areas for improvement. In summary, the successful implementation of industrial methodologies requires careful planning, clear communication, effective change management, and collaboration between change leaders and employees.

INTEGRATION OF INDUSTRIAL METHODOLOGIES INTO THE COMPANY'S STRATEGY

The modern industry is characterized by fierce competition in both national and international markets, the increasing complexity of production processes, and the intensive use of advanced technologies. To remain competitive and profitable, companies need to adopt a comprehensive strategy that addresses all aspects of their operation, from planning and design to production and customer delivery. In this context, the integration of industrial methodologies is a key element for business success. This chapter explores the main industrial methodologies currently used and their impact on business strategy.

Industrial Methodologies

There are numerous industrial methodologies used today to improve the efficiency, quality, and profitability of production processes. Some of the most popular methodologies include:

Lean Manufacturing: This methodology focuses on waste elimination and continuous improvement of production processes. It is based on five fundamental principles: value, flow, pull, perfection, and respect for people. Implementing Lean Manufacturing involves identifying and eliminating activities that do not add value to the process, improving workflow, eliminating bottlenecks, and implementing on-demand production processes.

Six Sigma: This methodology focuses on improving quality and reducing variability in production processes. It relies on data collection and analysis to identify and solve problems, reduce defects, and enhance customer satisfaction. Implementing Six Sigma involves clearly defining objectives, collecting and analyzing data, implementing solutions, and continuously measuring results.

Theory of Constraints: This methodology focuses on identifying and eliminating constraints that limit a company's production capacity. It is based on the idea that a production chain is only as strong as its weakest link. Implementing the Theory of Constraints involves identifying bottlenecks and implementing solutions to improve production capacity at the weakest link.

Total Productive Maintenance (TPM): This methodology focuses on maximizing the efficiency of production equipment through preventive maintenance and continuous improvement. It is based on the idea that regular maintenance and early problem identification can prevent costly failures and extend the lifespan of equipment. Implementing TPM involves identifying critical equipment, defining preventive maintenance procedures, training workers, and implementing a system to monitor and measure equipment efficiency.

Integration of Industrial Methodologies into the Company's Strategy

The successful implementation of industrial methodologies requires a comprehensive business strategy that addresses all aspects of the company's operation. Some key elements of a comprehensive strategy include:

Clear definition of business objectives

A comprehensive strategy should start with a clear definition of business objectives. Objectives should be Specific, Measurable, Achievable, Relevant, and Timely (SMART). Additionally, objectives should align with the company's vision and mission. Once objectives are defined, key performance indicators can be established to measure progress toward achieving those objectives.

Identification of critical improvement areas

Identifying critical improvement areas is a crucial step in integrating industrial methodologies into the company's strategy. Critical improvement areas may include reducing production costs, improving quality, reducing delivery times, enhancing customer satisfaction, and improving operational efficiency. Once critical improvement areas are identified, the most suitable industrial methodologies can be selected to address those issues.

Implementation of industrial methodologies

Implementing industrial methodologies involves defining the processes and procedures necessary to apply those methodologies in the company's operation. This may include training workers, defining work procedures, implementing specific tools and technologies, and measuring performance. It is essential to ensure that all methodologies are integrated consistently into the company's daily operation.

Monitoring and performance measurement

Monitoring and measuring performance are key elements of a comprehensive strategy for integrating industrial methodologies. Performance indicators should be established to measure progress toward business objectives and continuous improvement. Results should be regularly monitored and evaluated to ensure that business objectives are being achieved and to identify additional areas for improvement.

Culture of continuous improvement

The integration of industrial methodologies into the company's strategy must be accompanied by a culture of continuous improvement. Workers should be encouraged to constantly seek ways to improve the company's processes and procedures. Regular feedback and active worker participation are key elements in fostering a culture of continuous improvement.

Conclusion

The integration of industrial methodologies into the company's strategy is essential to remain competitive in the modern industry. Industrial methodologies can improve the efficiency, quality, and profitability of production processes and can be used to address a wide range of business problems. Successful implementation of industrial methodologies requires a comprehensive strategy that addresses all aspects of the company's operation, from defining objectives to fostering a culture of continuous improvement. By consistently integrating industrial methodologies into the company's daily operation, businesses can enhance their competitiveness and profitability in both national and international markets.

EXPERIENCES AND SUCCESS CASES IN THE IMPLEMENTATION OF INDUSTRIAL METHODOLOGIES

In the world of industry, the implementation of methodologies is a common practice aimed at improving the efficiency, productivity, and quality of production processes. However, carrying out this task is not always easy and can pose various challenges. In this chapter, we will explore some experiences and success cases in the implementation of industrial methodologies with the aim of better understanding the challenges and benefits of this process.

Experiences in the implementation of industrial methodologies:

To begin with, it is important to note that there is no one-size-fits-all methodology that works for all companies and situations. Each organization has its own needs, goals, and resources, so it is important to adapt the methodology to its specific context. Below are some experiences and common challenges in the implementation of industrial methodologies.

Experience 1: Lean Manufacturing Implementation

A metal parts manufacturing company decided to implement Lean Manufacturing methodology to improve the efficiency and productivity of its processes. The implementation took place in several phases, starting with the identification of critical processes and the elimination of waste. The Kanban system was also implemented to improve inventory management and reduce lead times.

One of the main challenges was employee resistance to change. Initially, many were reluctant to abandon their old ways of working and adapt to the new processes. However, through training and active involvement of workers in the implementation process, this barrier was overcome.

Another significant challenge was the measurement and tracking of results. Although a significant improvement in efficiency and productivity was observed, establishing a baseline and measuring the exact impact of Lean Manufacturing implementation on final results proved challenging. However, the company continued to make adjustments and improvements in the process, resulting in continuous improvement in outcomes.

Experience 2: Six Sigma Implementation

A telecommunications company decided to implement the Six Sigma methodology to improve the quality of its processes and reduce defects in its products. A dedicated team was formed to implement Six Sigma, responsible for identifying critical processes and analyzing data to identify the causes of problems.

One of the most important challenges was the lack of understanding of the methodology by employees. Many did not understand how Six Sigma worked and how they could contribute to the improvement process. To overcome this obstacle, a training and communication campaign was conducted to explain the methodology and its benefits.

Another challenge was the collection and analysis of data. The company found that it lacked the necessary systems to effectively collect data, so it had to invest in technology and data analysis tools. Once this issue was resolved, there was a significant improvement in product quality.

In addition to the mentioned challenges, the implementation of Six Sigma also brought significant benefits to the company. For example, the methodology helped the company standardize its processes, reducing variability and improving product consistency. It also identified hidden problems and addressed them before they affected product quality.

Another significant benefit was the improvement in customer satisfaction. By reducing defects in products, the company was able to deliver higher-quality products to its customers, improving their satisfaction and loyalty. This, in turn, had a positive impact on the company's reputation and financial success.

Experience 3: Total Productive Maintenance (TPM) Implementation

A food manufacturing company decided to implement TPM methodology to improve the efficiency and reliability of its production equipment. The implementation was divided into several phases, starting with the identification of critical equipment and conducting a failure analysis to determine the root causes of problems.

One of the most significant challenges was the lack of employee commitment to the methodology. Initially, many workers did not see the value in dedicating time and resources to TPM implementation and preferred to stick to their old ways of working. To overcome this obstacle, the company conducted a communication and training campaign to explain the methodology and its benefits to employees.

Another significant challenge was the lack of resources for TPM implementation. The company found that it did not have enough trained personnel to effectively carry out the implementation, so it had to invest in training and hire new employees to meet the needs.

Despite these challenges, TPM implementation had significant results for the company. Downtime of equipment was significantly reduced, and its reliability improved. Additionally, the efficiency of maintenance processes was improved, reducing costs and improving customer satisfaction by reducing delivery times.

Success Cases in the Implementation of Industrial Methodologies:

In addition to these experiences, there are several success cases in the implementation of industrial methodologies in different companies and sectors. Below are some of them:

Case 1: Toyota and the Implementation of Lean Manufacturing

Toyota is one of the most well-known examples of success in the implementation of Lean Manufacturing. The company has been a pioneer in the development and implementation of this methodology since the 1950s, significantly contributing to its success as an automobile manufacturer.

The implementation of Lean Manufacturing at Toyota is based on the following principles:

Waste elimination: The company focuses on identifying and eliminating any activity that does not add value to the production process.

Continuous improvement: Toyota constantly seeks to improve its processes and products, using customer and employee feedback to identify improvement opportunities.

Teamwork: The company encourages collaboration and teamwork among employees from different departments and hierarchical levels.

Just in time: Toyota's Lean Manufacturing methodology is based on the concept of just-in-time production, meaning that products are manufactured in the exact quantity and timing needed, without inventory accumulation.

Total quality: Toyota focuses on the total quality of its products, involving all employees in problem prevention and the identification of improvement opportunities.

Thanks to the implementation of Lean Manufacturing, Toyota has significantly reduced delivery times, improved the quality of its products, and reduced production costs. Additionally, the methodology has allowed the company to adapt quickly to market changes and maintain its position as a leader in the automotive industry.

Case 2: Johnson & Johnson and the Implementation of Six Sigma

Johnson & Johnson, a leading company in the health sector, decided to implement the Six Sigma methodology to enhance the quality and efficiency of its production processes. The company focused on implementing Six Sigma in its manufacturing processes, including the identification of critical processes, employee training, and the implementation of continuous improvement tools.

As a result of implementing Six Sigma, Johnson & Johnson significantly reduced defects in its products, improved process efficiency, and lowered production costs. Moreover, the methodology enabled the company to enhance customer satisfaction by delivering higher-quality products and reducing delivery times.

Case 3: GE and the Implementation of Total Quality Management

General Electric (GE) is a company recognized for its successful implementation of Total Quality Management (TQM). The TQM implementation at GE focused on continuous process improvement, waste elimination, and the involvement of all employees in problem prevention and the identification of improvement opportunities.

The implementation of TQM at GE had a significant impact on the company's efficiency and quality. For example, the company managed to significantly reduce product delivery times, improve product quality, and reduce production costs. Furthermore, the methodology allowed GE to enhance customer satisfaction and maintain its position as an industry leader.

Conclusion

The implementation of industrial methodologies can pose significant challenges, such as resistance to change and a lack of resources. However, it can also yield substantial benefits, including improved process and product efficiency, cost reduction, and enhanced customer satisfaction.

Through the experiences and success cases presented in this chapter, it is evident that the implementation of industrial methodologies can be a powerful tool for improving competitiveness and financial success across different sectors. To achieve successful implementation, strong commitment and effective communication with employees, along with investment in training and resources, are crucial to ensuring proper implementation and the continuation of continuous improvement.

Additionally, it is essential to recognize that there is no universally applicable methodology; each company must find the one that best suits its needs and characteristics. Careful evaluation of different options and adaptation to the company's culture and strategy are important considerations.

Finally, it is crucial to view the implementation of industrial methodologies not as an isolated project but as a continuous and evolving process. Continuous improvement should be ingrained in the company's culture and long-term strategy.

In summary, the implementation of industrial methodologies can be a powerful tool for improving competitiveness and financial success for companies. However, it is vital for each company to find the methodology that best suits its needs, and the implementation should be viewed as a continuous and evolving process.

CHALLENGES AND OPPORTUNITIES IN THE FUTURE OF INDUSTRIAL METHODOLOGIES

Industrial methodologies have experienced rapid advancement in recent decades, driven by the increasing demand for improvements in efficiency and quality in production processes. These methodologies, ranging from Six Sigma and Lean Manufacturing to Lean Production and Total Quality Management, have proven to be extremely effective in helping companies enhance their processes and increase profitability.

However, today, businesses face a series of increasingly complex challenges in the business environment. Globalization, digitization, and automation are changing the industrial landscape, creating new opportunities and challenges. This chapter will analyze the challenges and opportunities facing industrial methodologies in the future, presenting some innovative solutions to address them.

Challenges of Industrial Methodologies

Adaptation to the Digital Era

Digitization has transformed the way industrial processes are carried out, creating new opportunities for efficiency and optimization. However, many companies have not yet adapted to this new era. Incorporating new technologies, such as the Internet of Things (IoT) and artificial intelligence (AI), requires a significant investment in infrastructure and human resources. The lack of training and expertise can be an obstacle to the effective implementation of these technologies.

Flexibility in the Supply Chain

Globalization has allowed companies to access new markets and suppliers, but it has also created new complexities in the supply chain. Companies need to be able to adapt quickly to changes in demand and market conditions, requiring increased flexibility in the supply chain. Additionally, companies need to be prepared for inherent risks in the supply chain, such as natural disasters, political or social disruptions, and cyber-attacks.

Talent Management

The success of any industrial methodology largely depends on the talent and experience of the people implementing it. However, talent shortages in some sectors and regions can hinder the effective implementation of industrial methodologies. Moreover, the aging workforce and a lack of relevant skills in the younger workforce can limit companies' ability to implement new technologies and methodologies.

Sustainability

Companies are increasingly concerned about the environmental and social impact of their operations. Industrial methodologies can help companies reduce their environmental impact and improve sustainability. However, it is also essential to consider the long-term effects of processes and products. Sustainability also includes social considerations, such as working conditions and human rights throughout the supply chain.

Opportunities for Industrial Methodologies

Automation

Automation of industrial processes can provide increased efficiency and a reduction in human errors. Robotics and automation can help companies adapt to the growing demand for customization and variability in production, enhancing efficiency and reducing costs. Automation can also be a solution to talent shortages in some sectors, allowing companies to do more with fewer personnel.

Data Analytics

Data analytics can be a valuable tool for companies looking to improve efficiency and quality in production processes. Real-time data collection and analysis can provide insights into equipment performance, product quality, and process efficiency. This data can be used to identify areas for improvement and make informed decisions about the implementation of industrial methodologies.

Integrated Management Systems

Integrated management systems can help companies effectively manage production processes and the supply chain. These systems allow centralized management of processes, reducing complexity and improving efficiency. Additionally, integrated management systems can enhance communication between departments and suppliers, increasing transparency and reducing errors.

Sustainability

Sustainability can be an opportunity for companies looking to differentiate themselves in the market and meet the demands of increasingly environmentally conscious consumers. The implementation of industrial methodologies can help companies reduce their environmental impact and improve the sustainability of their operations. This may include waste and emission reduction, the use of renewable energy, and improved energy efficiency.

Innovative Solutions to Address Challenges

Training and Skill Development

Training and skill development can help companies address talent shortages and the lack of relevant skills in the workforce. Companies can invest in training programs to update workers' skills and prepare them for the implementation of new technologies and methodologies. Additionally, companies can collaborate with educational and governmental institutions to promote the development of skills relevant to the industry.

Staggered Implementation of Technologies

Staggered implementation of technologies can help companies tackle the challenge of adapting to the digital era. Instead of attempting to implement all technologies at once, companies can gradually implement technologies and assess their impact on processes. This approach can help companies reduce risk and initial investment, allowing them to adjust their strategies as they learn more about the technologies.

Supply Chain Collaboration

Collaboration in the supply chain can help companies improve the flexibility and resilience of their operations. Companies can collaborate with suppliers and customers to share information and plan jointly to reduce the risk of supply chain disruptions. Additionally, collaboration can help companies identify opportunities to improve efficiency and reduce costs throughout the supply chain.

Open Innovation

Open innovation can help companies develop innovative solutions to address current and future challenges. Open innovation involves collaborating with external partners, such as startups, universities, and other stakeholders in the business ecosystem, to develop new solutions and technologies. Open innovation can allow companies to access new ideas and skills, helping them stay at the forefront of innovation in their industry.

Use of Emerging Technologies

The use of emerging technologies, such as artificial intelligence, robotics, and augmented reality, can provide new solutions for current and future industrial challenges. These technologies can improve efficiency and accuracy in production, reduce downtime, and enhance worker safety. Additionally, these technologies can enable greater customization and variability in production, improving customer satisfaction and competitiveness.

Focus on Quality

A focus on quality can help companies improve customer satisfaction and efficiency in production processes. Implementing quality management systems, such as ISO 9001, can help companies establish clear and standardized processes to ensure the quality of products and services. Moreover, a focus on quality can help companies identify areas for improvement and reduce non-quality costs.

Conclusions

The future of industrial methodologies presents challenges and opportunities for companies. Adapting to the digital era, talent shortages, and the demand for sustainability are some of the challenges companies face. However, innovative solutions, such as training and skill development, staggered implementation of technologies, supply chain collaboration, open innovation, the use of emerging technologies, and a focus on quality, can help companies address these challenges.

Companies that can adapt to these challenges and leverage these opportunities will be better positioned to compete in an increasingly demanding and changing business environment. The implementation of industrial methodologies can help companies improve efficiency, quality, and sustainability in their operations, enhancing customer satisfaction and reducing costs. Furthermore, the implementation of industrial methodologies can help companies stay at the forefront of innovation in their industry and ensure long-term success.

CONCLUSIONS AND RECOMMENDATIONS FOR THE IMPLEMENTATION OF INDUSTRIAL METHODOLOGIES

The implementation of industrial methodologies is a key topic in the efficient management of companies. These methodologies have been developed over the years with the aim of optimizing production processes, reducing costs, and improving product quality. In this chapter, the main conclusions and recommendations for the implementation of industrial methodologies will be presented.

The implementation of industrial methodologies is essential for improving the efficiency of production processes in companies. These methodologies help reduce production times, enhance product quality, and lower costs.

One widely used methodology in the industry is Lean Manufacturing. This methodology focuses on eliminating waste in production processes and continuous improvement.

Another widely adopted methodology is Six Sigma, which aims to reduce variability in production processes. It identifies and corrects issues in processes, leading to improved product quality.

The Cleaner Production methodology is one of the newest in the industry, focusing on reducing the environmental impact of production processes through the implementation of cleaner and more sustainable practices.

Implementing industrial methodologies requires a systemic approach and active involvement of workers. It is crucial to engage all levels of the organization in the implementation process and promote a culture of continuous improvement.

Worker training and education are essential for implementing industrial methodologies. Employees need training in specific methodologies and continuous improvement of production processes.

Change management is a critical factor in implementing industrial methodologies. Effectively communicating the changes being made and ensuring workers understand and accept these changes are important.

Measuring and monitoring performance indicators are essential to evaluate the effectiveness of implemented industrial methodologies. Performance indicators must be clearly defined and aligned with the company's objectives.

Feedback is key to continuous improvement in production processes. Workers should be able to provide feedback on processes, and this feedback should be considered in the implementation of improvements.

Recommendations

Before implementing an industrial methodology, it is important to assess the specific needs of the company and determine the most suitable methodology to meet these needs.

Implementation of industrial methodologies should be led by a multidisciplinary team, including representatives from all levels of the organization, committed to continuous improvement.

Worker training and education are essential for implementing industrial methodologies. Workers must receive proper training to effectively implement and use methodologies. Training should be continuous and include both theory and practical implementation of methodologies.

Involving workers in the implementation of industrial methodologies is important. This can be achieved through effective communication of objectives and benefits, as well as participation in continuous improvement teams.

Change management is crucial for the success of industrial methodology implementation. Establishing a communication and change management plan to involve workers and facilitate the transition to new practices is important.

Defining performance indicators before implementing industrial methodologies is advisable. These indicators should be relevant, measurable, and aligned with the company's objectives. Additionally, establishing a plan for monitoring and evaluating performance indicators is important.

Feedback is a key component for continuous improvement in production processes. Establishing feedback channels for workers to provide comments and suggestions on production processes is important. This feedback should be considered in the implementation of improvements.

Implementing industrial methodologies is not a one-time process. It is important to establish a continuous improvement plan that allows the evolution of practices and the incorporation of new methodologies or practices for greater efficiency and quality in production processes.

Establishing strategic alliances with suppliers and customers to implement sustainable and continuous improvement practices throughout the supply chain is advisable.

Conclusion

The implementation of industrial methodologies is a key tool for improving the efficiency and quality of production processes in companies. Lean Manufacturing, Six Sigma, and Cleaner Production are some of the most widely used methodologies in the industry.

The implementation of these methodologies requires a systemic approach, active involvement of workers, continuous training and education, change management, definition of performance indicators, and constant feedback.

Companies should evaluate their specific needs and determine the most suitable methodology. It is essential to establish a continuous improvement plan for the evolution of practices and the incorporation of new methodologies for greater efficiency and quality in production processes.

In summary, the implementation of industrial methodologies is a continuous and dynamic process that requires commitment, collaboration, and active participation at all levels of the organization. Effective implementation of these methodologies can lead to significant improvements in the efficiency, quality, and sustainability of company production processes.

ABOUT THE AUTHOR

Industrial and Systems Engineer

Master in Administration with Quality and Productivity

3 Certifications

2 Technical Studies

More than 20 training courses

Winner of the Singapore Cooperation Programme ITE.

Instructor, engineer, content creator, and writer.
Discover the modern industry with the most controversial engineer.
engr's Workshop

I. Laisequilla

Author / Engineer

RELATED BOOKS

The bible of Industrial Engineering

From production management to process optimization, including methods and time engineering, this book covers the most important aspects of industrial engineering.

Available formats: physical, ebook, and audiobook

the all about Industrial Quality

FMEA, SPC, MSA, APQP, FMECA, Kaizen, Lean, ISO 9001, ISO 14001, ISO 45001, among others. Explained in an accessible and easy-to-understand manner.

Available formats: physical, ebook, and audiobook

the all about Lean Six Sigma

A complete and practical guide for those who wish to implement this methodology in their organization, with a detailed focus on the principles and theoretical foundations.

Available formats: physical, ebook, and audiobook